YOUR KNOWLEDGE HAS VALUE

- We will publish your bachelor's and
  master's thesis, essays and papers

- Your own eBook and book -
  sold worldwide in all relevant shops

- Earn money with each sale

Upload your text at www.GRIN.com
and publish for free

Rajni Garg

Supramolecular Chemistry of Host-Guest Inclusion Complexes

GRIN Publishing

Imprint:

Copyright © 2012 GRIN Verlag, Open Publishing GmbH
Print and binding: Books on Demand GmbH, Norderstedt Germany
ISBN: 978-3-656-13144-1

This book at GRIN:

http://www.grin.com/en/e-book/187894/supramolecular-chemistry-of-host-guest-inclusion-complexes

SUPRAMOLECULAR CHEMISTRY OF HOST-GUEST INCLUSION COMPLEXES

Supra-molecular chemistry is one of the most popular areas of experimental chemistry and it seems set to remain that way for the coming future. Not only the chemists, but biochemists, environmental scientists, engineers, physicists, theoreticians, mathematicians, and a lot of other researchers are also attracted towards this field of chemistry. Supra-molecular science thus had crossed traditional boundaries and more and more scientists are attracted into this field. The basic reason behind this interest is because of the presence of a number of supra-molecular assemblies in nature. [1-3]

The basic difference in the classical and supra-molecular chemistry is the difference in the bonds responsible for the connectivity of the molecules. In classical chemistry, molecular building blocks are connected by permanent covalent bonds. However, in supra-molecular chemistry, these bonds are replaced by molecularly matching physical bonds. These bonds can be hydrogen bonds to form liquid crystals, polymer-like chains, side chain-liquid crystalline polymers or comb copolymers. The consequences of these physical bonds include ionic complexation, steric and charge match in the case of crown ethers/metal cation complexes, π–π stackings, and coordination complexation. Due to these interactions, highly specific complexes called supra-molecules are formed, which are able to form a hierarchy of structures. These specific interactions are typical examples of molecular recognition. Molecular recognition is characterized by the simultaneous stability of the supra-molecule and selectivity in its formation. An important example of such case in the biological systems is the formation of the double-stranded structure of DNA and the binding of enzymes. [4-7]

Colloidal systems are the significant part of supra-molecular chemistry. On the one hand colloidal systems are extensively used in the application of detergents, emulsifiers/demulsifiers, foaming agents, wetting agents, etc., in solving the day-to-day problems that exist in many fields of industrial and domestic processing. On the other hand, these phenomena are also critical to the very fundamental processes of biological membrane formation and function in living cells. Colloidal science has generated a number of fields such as polymers, semiconductors, liquid crystals, membranes, vesicles and micellar aggregates. **Micellar aggregates** have served as an important bridge between microscopic and macroscopic chemical species in the development of new technologies. [8-9]

Materials exhibiting the characteristic of modifying the interfacial interactions by way of enhanced adsorptions are referred to as SURFace ACTive AgeNTS or **SURFACTANTS**. The term Surfactant was coined by Antara Products in 1950. Surfactants are characterized by the presence of two moieties in the same molecule viz. one polar head group (soluble in water) and other non-polar aliphatic tail (insoluble in water). Surfactants, therefore, can simultaneously dissolve in both hydrophobic and hydrophilic medium. [10-12]

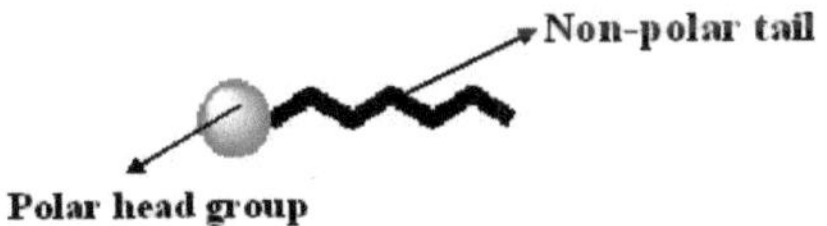

Structure of a surfactant monomer

Surfactants can be classified depending on the charge present in the hydrophilic group i.e. its headgroup (after dissociation in aqueous solution):

1. **Anionic surfactants** - The surfactants are called anionic, if the head groups are negatively charged such as sulphate, sulphonate or carboxylate anions. The most common example is Sodium dodecyl sulphate (SDS).

Sodium dodecyl sulfate (SDS)

2. **Cationic surfactants**- The surfactants are called cationic if the head groups are positively charged. The cationic surfactants are usually quaternary ammonium, imidazolinium or alkyl pyridinium compounds. For example- Cetyltrimethylammonium bromide (CTAB).

Cetyltrimethylammoniumbromide (CTAB)

3. **Zwitterionic surfactants-** The surfactants are called zwitterionic surfactants if the head groups contain both positive and negative charges. The particular example is Lecithin.

Lecithin

4. **Non-ionic surfactant -** The surfactants are called nonionic surfactants if the head group has no charge groups such as alkyl poly (ethylene oxide). For example- Polyoxyethylene (5) lauryl ether ($C_{12}E_5$).

Polyoxyethylene (5) lauryl ether

The existence of groups with opposing characteristics is responsible for all the special properties of surfactants. The properties of surfactants fall into two broad categories:

1) **Adsorption**: Adsorption is the property of surfactant molecules to collect at an interface (water / oil or water / air).

2) **Self-assembly**: Self-assembly is the tendency of a surfactant molecule to organize themselves into extended structures in water.

Upon increasing the concentration of the amphiphilic compound in water, at a certain point the solubility limit will be reached and phase separation will set in. Due to the efficient interactions between the polar headgroups and the surrounding water molecules, a complete phase separation is usually unfavourable. Instead, the process will be arrested in an intermediate stage with concomitant formation of aggregates of amphiphilic material, wherein the nonpolar parts stick together and are shielded from water, whereas the headgroups are located in the outer regions of the aggregate. A multitude of different aggregates can be formed in this way. The morphology of these assemblies is mainly determined by the shape of the individual surfactant molecules. These aggregates are called **micelles**. In non-polar medium, the micelle is inverted as compared to micelle in polar medium.

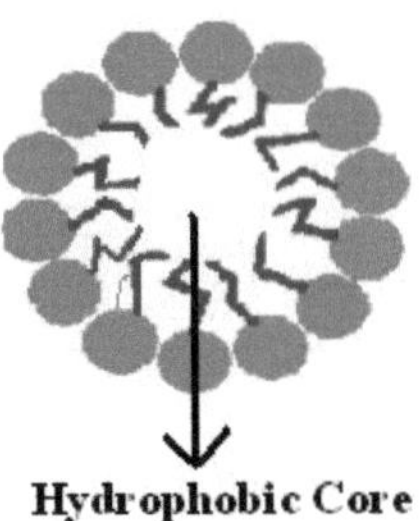

Hydrophobic Core

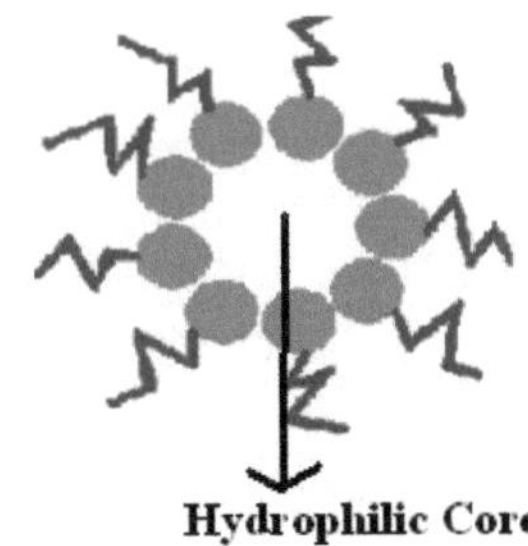

Hydrophilic Core

Structure of a micelle

The formation of micelles sets in after a certain critical concentration of surfactant (**the critical micelle concentration, cmc**) has been reached. Beyond this concentration the addition of more surfactant molecules will result in an increase in the number of micelles, while the concentration of monomeric surfactant remains almost constant. Micellisation is usually driven by an increase in entropy, resulting from the liberation of the water molecules from the hydrophobic hydration shells of the monomeric amphiphile molecules, whereas the enthalpy change is generally close to zero. [8-12]

Surfactants place a special role in modern day to day life and technological applications. Among the various types of surfactants, cationic surfactants are among the most extensively used surfactants in commercial products. Here the surface-active moiety has a positive charge, thus adsorbs strongly onto most solid surfaces and can impart special characteristics to the substrate, especially surfactants belonging to the class of long chain alkyl quaternary ammonium salts are very useful. They are unaffected by pH changes, positive charge remains in acidic, neutral and alkaline media. Generally they are used in textile softeners industrially and for home use in the rinse cycle of washing machines. They impart fluffy, soft "hand" to fabrics by absorbing on to them with hydrophobic group oriented away from fibers.

N-alkyltrimethyl ammonium chlorides are used as emulsifying agents for acidic emulsions or where adsorption of emulsifying agents on to substrate is desirable. Benzalkonium chloride is used as an antiseptic and spermicide. It is deemed safe for human use, and is widely used in eyewashes, hand and face washes, mouthwashes, spermicidal creams, and in various other cleaners, sanitizers, and disinfectants. Cetyltrimethylammonium bromide (CTAB) is one of the components of the antiseptic Cetrimide. Its uses include providing a buffer solution for the extraction of DNA. Lecithin is extracted from plant (soy) or eggs. There are very biocompatible. It's a natural component of cell membrane.

In anionic surfactants, Sodium dodecyl sulfate (SDS) is used in household products such as toothpastes, shampoos, shaving foams and bubble baths for its thickening effect and its ability to create lather. Sodium laureth sulfate, or sodium lauryl ether sulfate (SLES), is found in many personal care products (soaps, shampoos, toothpaste etc.). It is an inexpensive and very effective foamer. All of the soaps (sodium oleate, etc) are fatty acid salts. Docusate sodium [USP] is used as a stool softener and is administered orally or rectally as a tablet disintegrant or as an emulsifier and dispersant in topical preparations.

Recent reports suggest that various biological surfactants are also playing an important part in enhancing solubility of sparingly soluble substances. They can increase the solubility of a molecule in desired case via micellisation. Hydrophobic groups of surfactant form a micelle core which is liquid hydrocarbon-like in character, while their hydrated polar groups constitute a micelle outer shell in contact with water. They are capable of reducing the interfacial tension and contact angle between solid particles and aqueous media thus improving the drug wettability and increasing surface availability of the drug dissolution.

The solubilizing power of micelles is associated with the hydrocarbon core. Thus the overall influence on the drug solubility relates to the inclusion of poorly soluble drugs in the non-polar cavity of micelles. The enhancement in solubility is due to adsorption of surfactant at biological membranes and their subsequent penetration into the membrane, altering fluidity and increasing permeability. Hence, micellar solubilization is a powerful alternative not only for dissolving hydrophobic drugs in aqueous environments but also in various drug delivery and drug targeting systems where they minimize drug degradation and loss, prevent harmful side effects and increase drug bioavailability. Micellar systems can stay in blood long enough to provide gradual accumulation in the required area and their sizes permit them to accumulate in areas with leaky vasculature. [12-23]

Aqueous solution of surfactants often exhibit unusual physiochemical properties associated with the ordering of water molecules around the solute, as they distort water structure and therefore increase the free energy of the system. This process can be decreased among others by the aggregation of surfactants into micelles, which is also related to an increase of the apparent molar and partial molar properties. Above the cmc, sudden changes in many physicochemical properties have been observed in aqueous solution of surfactants.

The physical properties like surface tension, interfacial tension and detergency changes below the cmc with concentration but there is no change in these properties above cmc. Some other physical properties like density, equivalent conductivity show a change in slope below and above the cmc. Thus cmc represents a fundamental micellar quantity to study self-aggregation of amphiphilic molecules in solution and depends on a variety of factors including the number of carbon atoms, the structural arrangement of the hydrophile, presence of added electrolyte, pH and temperature. Factors that lower the cmc usually increase the lifetimes of the micelles as well as the residence times of the surfactant molecules in the micelle. Thus to obtain the precise values of cmc is of scientific interest. [24-29]

Micellar systems can also be characterized by structural parameters, such as a mean aggregation number, shape, and thermodynamic parameters. Due to the dynamic character, the size and shape of micelles are subject to appreciable structural fluctuations. Hence, micellar aggregates are polydisperse, as is demonstrated by small-angle neutron scattering data. Average aggregation numbers are typically in the range of 40-100. A variety of methods have been developed to study the structure, kinetics, and thermodynamics of surfactant micellar systems.

However, most techniques have focused on studying certain aspects of micelle structure and the kinetics of surfactant micellization. In recent years, new developments in thermodynamic studies have emphasized the quantitative interpretation of thermodynamic properties of surfactant solutions using various theoretical models of micellization. Modeling studies serve the useful purpose of enabling one to extract values of important parameters of the micellar systems with a minimum of experimental data. It has been revealed that the residence times of the individual surfactant molecules in the aggregate are typically in the order of 10^{-5}-10^{-6} seconds, whereas the lifetime of the micellar entity is about 10^{-3}-10^{-1} seconds. [8-12]

Micelles are extremely dynamic aggregates. Ultrasonic, temperature and pressure jump techniques have been employed to study the rate constants associated with the different equilibria involved. It has been concluded that the rates of uptake of monomers into micellar aggregates are close to diffusion controlled. The highly dynamic character of the micelles has for a long time successfully misled chemists in their conception of the structure of a micelle. Extensive discussions have focused on the conformation of the alkyl chains in the interior. Another topic of heated debate comprised the extent of water penetration into the hydrocarbon interior. Small-angle neutron scattering studies have resolved this matter by indicating that significant water penetration into the micellar core is unlikely. However, at the interface, extensive contact between water and the hydrocarbon chain segments definitely occurs. The headgroups of the micelle are extensively hydrated.

For ionic micelles, a large fraction of the counter-ions are located in the vicinity of the headgroups. These counter-ions normally retain their first hydration shell. The part of the surfactant that contains the headgroups and a variable fraction of the counter-ions is called the Stern region. This region comprises an appreciable electric field and a high concentration of ions (several molar) at the interface between the nonpolar interior and the aqueous exterior of the micelle and can be expected to exhibit unique properties. For pyridinium iodides the polarity of this region has been probed with the aid of the interionic charge transfer band characteristic for these species. The results indicate a somewhat reduced polarity of the Stern region compared to bulk water. [8-12]

Significant changes in micelle properties are related to changes in the energetic of interaction among the components in the micelle, systems of this type have been traditionally treated as mixed micelles, of which the primary interest is the surfactant-additive interactions. The properties of surfactants solutions largely depend upon the nature and amount of additives. Although the solubilization of additives (co-surfactants) in micelles makes the micellar system more complicated than the binary system, it provides an additional opportunity to explore micelle structure and micellar solution properties in terms of the interaction between the micelles and the additives.

The effect of additives on the properties of micellization in water-surfactant-additive ternary systems can be investigated using fundamental thermodynamic parameters for mixed micelle components, extracted through quantitative analysis of transfer functions of the additive from water to aqueous surfactant solution, in conjunction with models. The thermodynamic method has the advantage that it can evaluate the molar property changes in the system over a broad concentration range. These properties can be measured precisely, thus they are able to account for the effect of independent variables such as temperature and pressure, and they are readily amenable to mathematical modeling. It is one of the principal approaches to obtain additional information which can then be used to correlate the physicochemical properties of the surfactant-additive aqueous solution with the chemical structure of the components. [30-47]

Host–guest chemistry is a special category of Supra-molecular chemistry. Host-guest chemistry was defined by Cram in 1974. If one molecule is significantly larger than another and can wrap around it then it is termed the 'host' and the smaller molecule is its 'guest', which becomes enveloped by the host. Host molecules can be acyclic, macrocyclic, or oligomeric. It is a molecule which can non-covalently interact with a guest to form a host-guest complex. That is, there is a weak attractive interaction between the host and guest. Many cations, anions, and some neutral molecules can serve as guest compounds. Host–guest complexes include biological systems, such as enzymes and their substrates, with enzymes being the host and the substrates the guest. In terms of coordination chemistry, metal-ligand complexes can be thought of as host-guest species, where large (often macrocyclic) ligands act as hosts for metal cations. If the host possesses a permanent molecular cavity containing specific guest binding sites, then it will generally act as a host both in solution and in the solid state and there is a reasonable likelihood that the solution and solid state structures will be similar to one another. [48-54]

There are a number of Supra-molecular assemblies in nature. These biological assemblies often provide highly ordered microenvironments and specifically integrated functions. Enzymes and biomembranes are typical examples, which have survived in long-term evolution. These include a variety of organic and inorganic components such as proteins, cofactors, lipids, metal cations, ionophores and other functional elements. Their refined and elegant functions are beautifully controlled by precise molecular recognition between more than two active components. These components are specified and organized to work cooperatively in the protein or membrane matrix. Molecular recognition is an important term not only in natural but also in artificial systems. These discoveries have created Supra-molecular chemistry and its enormous diversity with enormous applications. [55-61]

Macrocycles are a versatile class of supramolecules. Macrocycles are important and powerful ligands, ubiquitous in transition metal coordination chemistry for the reasons that firstly they mimic important biological ligands developed long ago by nature, for example the porphyrin prosthetic group of many metalloproteins and secondly they impart thermodynamic and kinetic stabilities to their metal complexes uncommon or non-existent with ligands of less complex topology. For the purposes of this review, a macrocycle is defined as a cyclic compound with nine or more members including at least three donor atoms. The binding ability of macrocyclic hosts generally is greater than that of its acyclic analogue. This phenomenon called the "macrocyclic effect" is used primarily to describe the binding strength. The macrocyclic effect varies in magnitude and thermodynamic origin with different systems. The possible reason for this mechanism is the greater dipole-dipole repulsion in a macrocyclic compound compared with its acyclic analog.

The fact that macrocyclic complexes are involved in a number of fundamental biological systems has long been recognized. The importance of such complexes viz. to the mechanism of photosynthesis in plants or to the transport of oxygen in blood and respiratory systems motivated the scientists for the investigation of the metal ion chemistry of the systems in general. Schiff base macrocycles have been of paramount importance in macrocyclic chemistry. These were among the first artificial metal macrocyclic complexes to be synthesized. The metal complexes containing synthetic macrocyclic ligands have attracted a great deal of attention because they can be used as models for more intricate biological macrocyclic systems viz metalloporphyrins (hemoglobin, myoglobin, cytochromes, chlorophylls), corrins (vitamin B_{12}) and antibiotics (valinomycin, nonactin). [62-70]

Considerable progress in macrocyclic chemistry has been made in the past decade. The number of new macrocyclic ligands designed and synthesized has increased attention significantly. One reason for the interest in this field is that macrocyclic ligands have enhanced the promotion of selective binding and transformation of a large variety of substrates such as inorganic or organic cationic anionic species and neutral molecules as well as the role in determining the strength of the interactions between the macrocycle and guest molecule. Therefore, introduction in the molecular framework of structural features that impart high selectivity in the recognition of different guests is one of the goals in the design of synthetic macrocycles. [71-81]

An extensive structural, kinetic and thermodynamic study has been performed on the surfactants in various solvents including the effect of additives. Keeping in view the diversity and applicability of macrocycles, it is worthwhile to initiate the investigation of the effect of such compounds on the ion-solvent interactions. Macrocycles and surfactants can have strong interactions, which can affect both bulk and surface properties of the solutions. When macrocycles are used along with surfactants, due to the contemporary presence of hydrophilic and hydrophobic portions, they can interact with micelles both by hydrophobic and hydrophilic interaction. How the micellization in solution is affected by the addition of macrocycles can be an interesting area to investigate. Generalized interactions between macrocycles and surfactants are referred to as the 'binding' of surfactants by macrocycles, which may be site specific.

So far there is no sound base to elucidate the solution behavior of such mixtures where a rich pattern is obtained in the bulk solvent. It is also found that variation of temperature and concentration of both, surfactants and macrocycles influence their mutual interactions. Such system, in general can be analyzed for various quantities that characterize aggregates in the solution. The transport, volumetric, thermodynamic properties and Spectroscopic studies are of peculiar importance as they permit quantitative study of various phenomenons like the interactions and structural changes occurring in the medium. Hence a complete modeling can be coupled with the results of the various investigations to obtain a more detailed picture of the positioning of the surfactants within the macrocyclic cage. [82-118]

Several reports are available in literature on surfactant-macrocycle interactions. Komura et al have studied inclusion complexation of (11-ferrocenylundecyl) trimethylammonium bromide by cyclodextrin and its effects on electrochemical behavior of the surfactant. Electrochemical measurements on the binding of Sodium dodecyl sulfate and Dodecyltrimethylammonium bromide with Cyclodextrins has been studied by Yunus et al. Mwakibete et al has determined the complexation constants between alkyl pyridinium bromide and Cyclodextrins using electromotive force methods. Gharibi et al has studied the electrochemical studies of interaction between cationic surfactant, Cetyltrimethylammonium bromide and Cyclodextrins at various temperatures. The thermodynamic behavior of the ternary systems comprising Cyclodextrin, Sodium octanoate or Sodium decanoate and water has been studied from density and speed of sound measurements by Tardajos et al. Apparent molar volumes and heat capacities of the two surfactants, sodium dodecanoate /sodium perfluorooctanoate mixtures in the water/cyclodextrin solvent as functions of the surfactant total molality were determined by Lisi et al. [71-76]

Volumetric study of modified Cyclodextrin/Hydrocarbon and Fluorocarbon surfactant inclusion complexes in aqueous solutions has been done by Wilson et al. Junquera et al have studied the effect of the presence of Cyclodextrin on the micellization process of Sodium dodecyl sulfate or Sodium perfluorooctanoate in water. Alfonso et al have investigated the mechanisms governing the interaction of equimolecular mixtures of sodium dodecyl sulfate (SDS) with phosphatidylcholine liposomes. Tardajos et al have studied the inclusional process of Decyltrimethylammonium bromide (DTAB) into the cavity of Cyclodextrin measuring speed of sound. [77-79]

Tong et al have used a numerical method based on surface tension measurements for determination of association constants for Cyclodextrin-Surfactant inclusion complexes. The surfactant behavior of the water-soluble calix[4]arene based amphiphiles and its complexation to Cyclodextrin were studied by Michels et al. Bojinova et al have done the physicochemical studies on complexes between Cyclodextrin and aliphatic guests as new noncovalent amphiphiles. Silva et al have done the implications of micellization in stoichiometry determination and model calculations on aqueous solution inclusion of the nonionic surfactant in cyclodextrin.

Rio et al have also carried out a study of the spectral behavior of crystal violet in mixed systems formed by Hexadecyltrimethylammonium chloride and Cyclodextrin. (80-84)

Datta et al have used fluorescence probe to investigate the interactions of cyclodextrin and neutral surfactants. Tardajos et al have carried NMR study on inclusion complexes between Cyclodextrin and a gemini surfactant in Aqueous Solution. Micellar aggregation numbers have been obtained for Sodium dodecyl sulfate micelles in cyclodextrin solutions through static fluorescence quenching by Obe et al. Bertorelle et al has compared behavior of hydrophobic fluorescent NBD probes in micelles and in Cyclodextrins. (85-88)

Park and Song have done fluorescence-probed studies on the 1:1 and 1:2 complexation of anionic surfactants with Cyclodextrin. Nicholas et al has done analysis of static and dynamic host-guest associations of detergents with Cyclodextrins via photoluminescence methods. Wagner et al have done a visual demonstration of Supra-molecular chemistry by observable fluorescence enhancement upon host–guest inclusion. A spectroscopic study on binding behaviors of different structural nonionic surfactants with Cyclodextrins has been carried out by Chen et al. (89-92)

The number of macrocyclic compounds has increased extensively over the post decades. An impressive number of new macrocyclic families of natural, seminatural, or artificial compounds like Porphyrins, Cyclodextrins, Calixarenes, Coronands, Cryptands and Crown ethers etc. have come into light. Out of these compounds, Crown ethers are of particular interest. **Crown ethers** are multidentate macrocyclic compounds, so called because of their appearance of space-filling models and their ability to "crown" cations. Crown ethers were discovered by Pederson in 1967. The term "crown" refers to the resemblance between the structure of crown ether bound to a cation, and a crown sitting on a head. Crown ethers belong to the category of macrocyclic compounds which undergo predominantly hydrophilic interaction and interact mainly with a wide variety of cations. The most significant property of crown ethers is to form inclusion complexes with different kinds of guest molecules, ranging from ionic and neutral to organic or inorganic chemical species.

Crown ethers are heterocyclic chemical compounds that consist of a ring containing several ether groups. The most common crown ethers are oligomers of ethylene oxide, the repeating unit being ethyleneoxy, i.e., $-CH_2CH_2O-$. Important members of this series are the tetramer (n = 4), the pentamer (n = 5), and the hexamer (n = 6). The first number in crown ether's name refers to the number of atoms in the cycle, and the second number refers to the number of those atoms that are oxygen. Thus, 15-crown-5 is comprised of 15 atoms in the ring, 5 of which are O and 10 of which are C. Structures of three typical crown ethers are:

18-crown- 6   15-crown-5

The complementarity or size-match between the metal ion and macrocycles ensures formation of the most stable complex. If the diameter of the cation matches the size and shape of the cavity, the cation can be bound tightly within the cavity. In crown ethers, each of the oxygen atoms possesses a lone electron pair that points inward, toward the cavity. This arrangement recounts the cavity to attract and to solvate cations. If the lone pairs are directed outward, the inside of the cavity should be nonpolar; the crown ether cannot complex with the cations in this case.

The outside of the crown ether consists mostly of nonpolar C-H bonds; this region of the crown ether should associate with other nonpolar molecules and should display larger affinity toward nonpolar solvents. A classic example of size-match selectivity is proved by crown ethers like, 14-crown-4 hosts selectively bind to Li^+, 15-crown-5 hosts selectively bind to Na^+, and 18-crown-6 systems complex selectively with K^+. However, one crown ether can complex with different metal cations. [120-121]

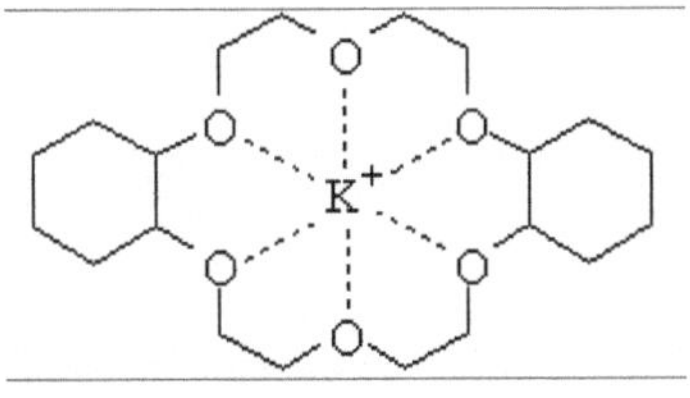

A host-guest complex

Crown ethers and related macrocycles are known to mimic some parts of biological molecular recognition and to mediate subsequent chemical processes. Since crown ethers bring several ions and/or molecules together around the crown ether rings, they can catalyze reactions in an enzyme mimetic manner and facilitate membrane transport, as do biological ionophores. Artificial Supra-molecular assemblies may have advantages over biological assemblies: facile synthesis, high physical stability and versatile molecular structure. Some crown ethers contain special groups, which renders them practically suitable for use as chemical sensors.

In the medical applications field, crown ethers are used in clinical analysis as diagnostic agents and as therapeutic agents. A major application of crown ethers in this area involves the fabrication of ion selective electrodes for medical use. Crown ethers have been used as diagnostic agents in the human body to help locate a tumor site. Crown ethers also can be useful as administered therapeutic agents to help remove toxins from the human body. Among the various properties of crown ethers, their unique chemical architecture plays a prominent role and opens the way to practical design of host molecules for selective complexation of various metal ions.

Although most artificial reported systems still exhibited functions inferior to those of corresponding biological systems, Scientists have opened the door to the Supra-molecular world in which artificial Supra-molecular demonstrate potentialities comparable with nature's evolutionary systems. Thus, the next goal is the establishment of general working principles for the design of Supra-molecular function based on specific recognition. Since crown ether chemistry provides useful strategies for molecular recognition and its combination with molecular assembly technology offers many opportunities in this new and fascinating field. Design, synthesis and functions of a new crown ether family applicable for construction of a

Supra-molecular assembly have already been described. Some interesting approaches towards supra-molecular assembly and functionalization have been reported.

Recent advances in host-guest chemistry demand more and more sophisticated and specific artificial host molecules. In order to realize such molecules, the availability of the appropriate synthetic methods of the host themselves or their intermediate is of primary importance. Crown ethers are frequently employed as phase transfer catalysts to extract metal ions from the aqueous phase into the organic phase. Several reactions are facilitated when they are carried out in the presence of a phase transfer catalysts. Crown ethers can catalyze reactions by transporting otherwise insoluble reactants into organic phases to participate in the reaction of interest.

Several investigations based on crown ethers in micellar systems have been reported in the literature. E. Vikingstad and J. Bakken have investigated the effect of complex-forming crown ethers, on micelle formation of sodium decanoate at 25°C by ultrasound measurements. Variations of cloud point have been studied in mixed anionic- nonionic surfactant solutions following the complexation of the sodium counter-ions and cryptand-222 by L. Marszall. Pedone et al have performed a small-angle neutron scattering study at fixed surfactant content and varying macrocycle concentrations to give information on its distribution between micellar and aqueous phases. They have also investigated partition of crown ethers in sodium dodecyl sulfate aqueous solutions and their effect on the structure of surfactant aggregates. In further studies, the structure of aggregates of a new bolaform surfactant in neat aqueous solution was also investigated by small-angle neutron scattering. [122-126]

Cesium and potassium dodecyl sulfate micellar solutions have been studied in the presence of three calix[4]arene-crown ethers able to complex cesium and potassium micellar counter-ions in a selective way by Capuzzi et al. The effect of 18-crown-6 on the decomposition of N-methyl-N-nitroso-p-toluenesulfonamide by N_3^- and H_3O^+ in micelles of tetradecyltrimethylammonium azide and hydrogen dodecyl sulfate has been studied by Agra et al. Floriano et al have studied the effects of the addition of crown ether on the structure of aqueous solutions of surfactants sodium dodecyl sulfate and dodecyltrimethylammonium bromide by small angle neutron scattering. [127-129]

Bencini et al studied the effects of addition of the macrocyclic cage CESTO on the aggregation behavior of Li dodecyl sulfate in aqueous solution. Gambi et al studied the properties of LDS and SDS micellar solutions upon the addition of C222 by small-angle neutron scattering. Teixeira et al have studied the effect of macrocyclic ligands on the aggregation behavior of lithium dodecyl sulfate micellar solutions. The study was extended further to determine the complexation of the Li^+ ion by macrobicyclicdiazapolyoxa ligands in a aluminum (III) chloride-N-(n-butyl) pyridinium chloride mixture at 40°C. [130-132]

Micellar solutions of the nonionic surfactant octyl-18-crown-6 have been studied through SANS as a function of the surfactant concentration with and without salt addition by Nostro et al. Counterion complexation by the cryptand in micellar solutions of lithium dodecyl sulfate was studied by M. Ginley and U. Henriksson. Selective encapsulation of lithium ions by CESTO has been studied by   Baglioni et al in the presence of micellar solutions. [135-137]

SANS experiments were carried out in order to determine the structure of the micellar system, and the interactions between the ligand and the micellar interface by Capuzzi et al. Resonance spectra of photogenerated N, N," N'-tetramethylbenzidine cation radical and n-doxylstearic acids in frozen micellar solutions of sodium and lithium dodecyl sulfate containing 15-crown-5 and 18-crown-6 ethers in D_2O have been studied as a function of crown ether concentration by Kevan et al. Electron Spin Resonance and Proton Matrix Electron Nuclear Double Resonance Studies of N,N,N',N'-Tetramethylbenzidine Photoionization in Sodium and Lithium dodecyl sulfate micelles were carried out by Hugh et al to study structural effects of crown ethers. Carime investigated Host-guest effect using Chorand crown ether macrocycle–alkali met al cation 1:1 complexes directly by mass spectrometry. [136-139]

P. Baglioni and L. Kevan studied structural effects of alcohol and crown ether addition by Electron spin resonance and electron spin echo modulation studies of N,N,N',N'-tetramethylbenzidine photoionization in sodium dodecylsulfate micelles. Ginley et al studied the effect of counterion complexation on micellar structure and dynamics by a NMR relaxation and self-diffusion study. Myassoedova et al studied the concentration effect of CE addition on the interfacial structure of sodium lauryl sulfate micelles and dihexadecyl phosphate vesicles, through two electron transfer. Quenching of pyrene fluorescence by Cesium ions in micellar systems was studied by Turro et al. Kuo et al carried a fluorescence probe investigation of the effect of alkali metal ions on the micellar properties of a crown ether surfactant. [140-146]

Computer modeling studies were undertaken in nonionic alkylpolyglycoside solutions containing cyclodextrin by V.C. Reinsborough and V.C. Stephenson in order to study the host-guest interactions. J.C. Schulz and G.G. Warr studied selective flotation of ions by macrocyclic complexation. E. M. Elnemma and S.R. Salman studied the effect of surfactant on the charge transfer complexes of 18C6 with Picric acid in the presence of alkali metal ions. Warr et al studied a comparison of counterion effects in surfactant and classical colloid systems on addition of macrocyclic compounds. Kaifer et al studied drastic decrease of the critical micelle concentration of Sodium dodecyl and Sodium decyl sulfates on addition of Cryptand-222. [147-151]

The rational design of new formulations requires a good knowledge of the encapsulation process. Structural information, such as the stoichiometry and the geometry of the complex, and thermodynamic information of binding, are necessary to draw a complete picture of the driving forces governing the crown ether- surfactant interactions.

REFERENCES

1. J. M. Lehn, Science, 1993, 260, 1762.

2. J. M. Lehn, Supramolecular Chemistry, Wiley, 1995.

3. G. V. Oshovsky, D. N. Reinhoudt and W. Verboom, Angew. Chem. Int. Ed. 2007, 46 (14), 2366.

4. J. M. Lehn, Angew. Chem. Int. Ed., 1990, 29 (11), 1304.

5. D. Philp and J. F. Stoddart, Angew. Chem. Int. Ed. Engl. 1996, 35, 1154.

6. S. R. Batten and R. Robson, Angew. Chem. Int. Ed. 1998, 37, 1460.

7. H. Dodziuk, Introduction to Supramolecular Chemistry, 2002, 43.

8. P. Somasundaran, Encyclopedia of Surface and Colloid Science, 2nd edition, 2006.

9. M. J. Rosen, Surfactant and Interfacial Phenomena, 1989.

10. T. Cosgrove, Colloid Science: Principles, Methods and Applications, 2005.

11. U. Tsoler and G. Broze, Handbook of Detergents: Properties, 1st edition, 1999.

12. D. Myers, Surfactant Science and Technology, 3rd Edition, 2005.

13. M. L. Gonzalez-Martin, B. Janczuk, J. A. Mendez-Sierra and J. M. Bruque, Colloids Surf. A, 1999, 148, 213.

14. A. Gonzalez-Perez, J.L. Delcastillo, J.Czapkiewicz and J.R. Rodriguez, Colloids Surf. A, 2004, 232, 183.

15. M. Perz, L.M. Varela, P. Taboada, D. Attwood and V. Mosquera, Colloid. Polym. Sci., 2000, 278, 706.

16. M. Fujiwara, T. Okano, T.H. Nakashima, A.A. Nakashima and G. Sugihara, Colloid. Ploym. Sci., 1997, 275, 474.

17. R.J. Williams, J.N. Phillips and K.J. Mysels, Trans. Faraday. Soc., 1955, 51, 561.

18. A. Gonzalez-Perez, J.L. Delcastillo, J. Czapkiewicz and J.R. Rodriguez, Colloid. Polym. Sci., 2002, 280, 503.

19. A. Gonzalez-Perez, J. Czapkiewicz, J.L. Delcastillo and J.R. Rodriguez, Colloid. Ploym. Sci., 2003, 281, 5561.

20. A. Gonzalez-Perez, J. Czapkiewicz, J.L. Delcastillo and J.R. Rodriguez, Colloids Surf. A, 2001, 193, 129.

21. A. Gonzalez-Perez, J. Czapkiewicz, J.L. Delcastillo and J.R. Rodriguez, J. Phys. Chem. B , 2001, 105, 1720.

22. J.R. Rodriguez, A. Gonzalez-Perez, J.L. Delcastillo and J. Czapkiewicz, J. Colloid. Interface. Sci., 2002, 250, 438.

23. K.H. Kang, H.U. Kim and K.H. Lim, Colloids Surf. A, 2001, 189, 113.

24. E. Junquera, J.C. Romero and E. Aicart, Langmuir, 2001, 17, 1826.

25. G. Gonzalez, A. Compostizo, L.S. Martin and G. Tardajos, Langmuir, 1997, 13, 2235.

26. B.D. Wagner, N. Stojanovic, G. Leclair and C.K. Jankowski, J. Incl Phenom. Macro. Chem., 2003, 45, 275.

27. C. Merino, E. Junquera, J. Barbera and E. Aicart, Langmuir, 2000, 16, 1557.

28. D.M. Davies and J.R. Savage, J. Chem. Soc. Perkin Trans., 1994, 2, 1525.

29. M.E. Brewster, R. Vandecruys, G. Verreck, M. Noppe and J. Peeters, J. Incl. Phenom. Macro. Chem., 2002, 44, 35.

30. A.P. Rossel, Quim. Nova, 2000, 6, 23.

31. E. Junquera, L. Pena and E. Aicart, J. Pharm. Sci., 1998, 87, 86.

32. A. Qi, L. Li and Y. Liu, J. Incl Phenom. Macro. Chem., 2003, 45, 69.

33. H. Ueda, J. Incl. Phenom. Macro. Chem., 2002, 44, 53.

34. N. Martinez, E. Junquera and E. Aicart, Phys. Chem. Chem. Phys., 1999, 1, 4811.

35. K.P.R. Chowdhary and T. Manjula, Ind. J. Pharm. Sci., 2000, 62 (2), 97.

36. G. Gonzalez-Gaitano, A. Crespo, and G.Tardajos, J. Phys. Chem. B, 2000, 104, 1869.

37. E. Junquera, G. Tardajos and E. Aicart, Langmuir, 1993, 9, 1213.

38. S. Schmolzer and H. Hoffmann, Colloids Surf. A, 2003, 213, 157.

39. K. Uekama, J. Incl. Phenom. Macro. Chem., 2002, 44, 3.

40. A. Stefansson and T. Loftssonn, J. Incl. Phenom. Macro. Chem., 2002, 44, 23.

41. M. Perez-Rodriguez, L. M. Varela, P. Taboada, D. Attwood and V. Mosquera, Colloid Polym. Sci., 2000, 273,706.

42. L. Seoane, P. Martinez, L. Besada, J.M. Ruso, F. Sarmiento and G. Prieto, Colloid Polym. Sci., 2002, 28, 624.

43. P. Taboada, J.M. Ruso, M. Garcia and V. Mosquera, Colloid Polym. Sci., 2001, 279, 716

44. M. Gautierrez, S. Barbosa, P. Taboada and V. Mosquera, Colloid Polym. Sci., 2003, 281, 575

45. M. Alauddin and R. E. Verrall, J. Phys. Chem., 1986, 90, 1647.

46. M. Alauddin and Ronald E. Verrall, J. Phys. Chem., 1989, 93, 3724.

47. L. Wang and R. E. Verrall, J. Phys. Chem., 1994, 98, 4368.

48. C.J. Pedersen, J. Inclusion Phenom., 1988, 6, 337.

49. D.J. Cram, J.M. Cram, Science, 1974, 183, 803.

50. D.J. Cram, Angew. Chem., Int. Ed. Engl., 1986, 25, 1039.

51. R.M. Izatt, J.S. Bradshaw, S.A. Nielsen, J.D. Lamb, J.J. Christensen and D. Sen, Chem. Rev., 1985, 85, 271.

52. F.P. Schmidtchen and M. Berger, Chem. Rev., 1997, 97, 1609.

53. F. Vogtle, H. Sieger and W.M. Muller, Top. Curr. Chem., 1981, 98, 107.

54. J.W. Steed, P.C. Junk and B.J. McCool, J. Chem. Soc., Dalton Trans., 1998, 3417.

55. H. Hassaballa, J.W. Steed, P.C. Junk and M.R. Elsegood, J. Inorg. Chem., 1998, 37, 4666.

56. P.D. Orince and J.W. Steed, Supramol. Chem., 1998, 10, 155.

57. E. Weber and F. Vogtle, Top. Curr. Chem., 1981, 98, 1.

58. J.C. Ma and D.A. Dougherty, Chem. Rev., 1997, 97, 1303.

59. P.D. Beer, M.G.B. Drew, P.A. Gale, P.B. Leeson and M.I. Ogden, J. Chem. Soc., Dalton Trans., 1994, 3479.

60. D.A. Hoic, M. Dimare, and G.C. Fu, J. Am. Chem. Soc., 1997, 119, 7155.

61. F. Paul, D. Carmichael, L. Ricard, and F. Mathey, Angew. Chem., Int. Ed. Engl., 1996, 35, 1125.

62. D.K. Cabbiness and D.W. Margerum, J. Am. Chem. Soc., 1969, 91, 6540.

63. D.K. Cabbiness and D.W. Margerum, J. Am. Chem. Soc., 1970, 92, 2151.

64. R.D. Hancock and A.E. Martell, Comments Inorg. Chem., 1988, 6, 237.

65. B.L. Haymore, R.M. Lamb, R.M. Izatt, and J. Christensen, J. Inorg. Chem., 1982, 21, 1598.

66. H.K. Frensdorff, J. Am. Chem. Soc., 1971, 93, 600.

67. F.P. Hinz and D.W. Margerum, Inorg. Chem., 1974, 13, 2941.

68. L. Fabbrizzi, P. Paoletti, and A.B.P. Lever, Inorg. Chem., 1976, 15, 1502.

69. M. Kodama, and E. Kimura, J. Chem. Soc., Dalton Trans., 1976, 2341.

70. F. Marsicano, R.D. Hancock, and A. McGowan, J. Coord. Chem., 1992, 25, 85.

71. T. Komura, T. Yamaguchi, K. Noda, and S. Hayashi, Electrochim. Acta, 2002, 47, 3315.

72. W.M.Z. Wan-Yunus, J. Taylor, D.M. Bloor, D.G. Hall, and E. Wyn-Jones, J. Phys. Chem., 1992, 96, 8979.

73. H. Mwakibete, D.M. Bloor and E. Wyn-Jones, Langmuir, 1994, 10, 3328.

74. H. Gharibi, S. Jalili and T. Rajabi, Colloids Surf. A, 2000, 175, 361.

75. G. Gonzalez-Gaitano, T. Sanz-Garcia, and G. Tardajos, Langmuir, 1999, 15 (23), 7963.

76. R. De-Lisi, G. Lazzara, S. Milioto and N. Muratore, Phys. Chem. Chem. Phys., 2003, 5, 5084.

77. L. D. Wilson and R. E. Verrall, J. Phys. Chem. B, 1998, 102 (2), 480.

78. A. de-la-Maza and J. L. Parra, Langmuir, 1996, 12, 3393.

79. E. Junquera, E. Aicart and G. Tardajos, J. Phys. Chem., 1992, 96(11), 4533.

80. R. Lu, J. Hao, H. Wang and Linhui Tong, J. Colloid. Interface. Sci., 1997, 192, 37.

81. J.J. Michels, J. Huskens, J.F.J. Engbersen, and D.N. Reinhoudt, Langmuir, 2000, 16, 4864.

82. T. Bojinova, Y. Coppel, N.L. Viguerie, A. Milius, I.R. Lattes, and A. Lattes, Langmuir, 2003, 19, 5233.

83. L. Silva and J.C. Teixeira, J. Incl. Phenom. Macro. Chem., 2002, 43, 127.

84. L.G. Río and A. Godoy, Phys. Chem. B, 2007, 111, 6400.

85. P.P. Mishra, R. Adhikary, P. Lahiri and A. Datta, Photochem. Photobiol. Sci., 2006, 5, 741.

86. A.G. Martinez, G.G. Gaitano, M.H. Vinas, and G. Tardajos, J. Phys. Chem. B, 2006, 110 (28), 13819.

87. D.J. Jobe, V.C. Reinsborough and S.D. Wetmore, Langmuir, 1995, 11(7), 2476.

88. A. Bertorelle, R. Dondon, and S. F.Forgues, J. Fluor., 2002, 12(2), 205.

89. J.W. Park and H.J. Song, J. Phys. Chem. 1989, 93, 6454.

90. J. Nicholas, T. Okubo, and C.J. Chung, J. Am. Chem. Soc., 1982, 104 (7), 1789.

91. B.D. Wagner, P.J. Macdonald, and M. Wagner, J. Chem. Edu. 2000, 77 (2)178.

92. X. Du, X. Chen, W. Lu, and J. Hou, J. Colloid Interface Sci., 2004, 274, 645.

93. D. Monti, M. Venanzi, V. Cantonetti, S. Boroccib and G. Mancini, Chem. Commun., 2002, 774.

94. D. Monti, M. Venanzi, V. Cantonetti, S.F. Ceccacci, C. Bombelli and G. Mancini, Chem. Commun., 2004, 972.

95. M. Nicolle and A.E. Merbach, Chem. Commun., 2004, 85.

96. O.K. Medhi and J. Silver, J. Chem. Soc., Chem. Commun., 1989, 1199.

97. E.B. Caruso, E. Cicciarella and S. Sortino, Chem. Commun., 2007, 5028.

98. E. Mileo, P. Franchi, R. Gotti, C. Bendazzoli, E. Mezzina and M. Lucarini, Chem. Commun., 2008, 1311.

99. L. Guo and Y. Liang, Supramol. Chem., 2004, 16 (1), 31.

100. C.M. Gandini, V.E. Yushmanov, I.E. Borissevitch, and M. Tabak, Langmuir, 1999, 15, 6233.

101. L. Guo, J. Colloid Interface Sci., 2008, 322 (1), 281.

102. B.M. Kurdziel , K. Solymosi , J. Kruk , B. Boddi and K. Strzałka , J. Photochem. Photobiol. B, 2007, 86 (3), 262.

103. L.P. Aggarwal and I.E. Borissevitch, Spectrochim. Acta A, 2006, 63 (1), 227.

104. P.S. Santiago, S. Neto, S.C. Gandini and M. Tabak, Colloids Surf. B, 2008, 65 (2), 247.

105. N.C. Maiti, S. Mazumdar, and N. Periasamy, J. Phys. Chem. B, 1998, 102 (9), 1528.

106. W.G. Qiu , Z.F. Li , G.M. Bai , S.N. Meng , H.X. Dai and H. He, Spectrochim. Acta A, 2007, 68 (5), 1164.

107. X. Li , D. Li , W. Zeng , G. Zou and Z. Chen, J. Phys. Chem. B., 2007, 111 (6), 1502.

108. Y.T. Wang and W.J. Jin, Spectrochim. Acta, Part A, 2008, 70 (4), 871.

109. L. Guo and Y. Q. Liang, Spectrochim. Acta, Part A, 2003, 59 (2), 219.

110. X. Li, Z. Zheng, M. Han, Z. Chen and G. Zou, J. Phys. Chem. B, 2007, 111 (17), 4342.

111. D.K. Das, O.K. Medhi , J. Inorg. Biochem., 1998, 70 (2), 83.

112. X. Li, Y. Xie, Z. Chen, G. Zou, Spectrochim. Acta, Part A, 2005, 61 (11), 2468.

113. A.A. Sivash, Z. Masinovsky, G.I. Lozovaya, Biosystems, 1991, 25 (3), 131.

114. E. Junquera, L. Pena, and E. Aicart, Langmuir, 1997, 13 (2), 219.

115. M. Vermathen, E.A. Louie, A.B. Chodosh, S. Ried, and U. Simonis, Langmuir, 2000, 16, 210.

116. S.C.M. Gandini, R. Itri, D. S. Neto, and M. Tabak, J. Phys. Chem. B, 2005, 109 (47), 22264.

117. P.S. Santiago, D.S. Neto, L.R. Barbosa, R. Itri and M. Tabak, J. Colloid Interface. Sci., 2007, 316 (2), 30.

118. S.C. Gandini , V.E. Yushmanov and M. Tabak, J. Inorg. Biochem., 2001, 85 (4), 263.

119. J.L. Atwood and J.W. Steed, Encyclopedia of supramolecular chemistry, 2004.

120. C.J. Pedersen, J. Am. Chem. Soc., 1967, 89, 2495, 7017.

121. M. Ginley and U. Henriksson, J. Colloid Interface Sci., 1992, 150 (1), 281.

122. E. Vikingstad and J. Bakken, J. Colloid Interface Sci., 1980, 74 (1), 8.

123. L. Marszall, Colloids and Surfaces, 1989, 35 (1), 1.

124. D.C. Martino, E. Caponetti and L. Pedone, Langmuir, 2004, 20 (10), 3854.

125. D.C. Martino, E. Caponetti and L. Pedone, J. Appl. Cryst., 2003, 36, 562.

126. E. Caponetti, D.C. Martino and L. Pedone, J. Appl. Cryst., 2003, 36, 753.

127. G. Capuzzi, E. Fratini, F. Pini, P. Baglioni, A. Casnati, and J. Teixeira, Langmuir, 2000, 16 (1), 188.

128. C. Agra, S. Amado, J.R. Leis, and A. Ríos, J. Phys. Chem. B, 1997, 101 (39), 7780.

129. E. Caponetti, D.C. Martino, M.A. Floriano, R. Triolo and G.D. Wignall, Langmuir, 1995, 11 (7), 2464.

130. P. Baglioni, A. Bencini, L. Dei, C.M. C.Gambi, P.L. Nostro, S.H. Chen, Y.C. Liu, J. Teixeira and L. Kevan, J. Phys.: Condens. Matter, 1994, 6 (23A), 369.

131. P. Baglioni, C.M.C. Gambi, R. Giordano and J. Teixeira, Colloids Surf. A, 1997, 121 (1), 10, 47.

132. P. Baglioni, Y.C. Liu, S.H. Chen and J. Teixeira, J. Phys. IV, 199, 3, 169.

133. P. Baglioni, C.M. C. Gambi, R. Giordano and J. Teixeira, Physica B, 1995, 213, 597.

134. R.R. Rhinebarger and A.I. Popov, Polyhedron, 1988, 7 (15), 1341.

135. P. Baglioni, C.M.C. Gambi, R. Giordano, J. Teixeira and P.L. Nostro, Physica B, 1997, 234, 300.

136. G. Capuzzi, E. Fratini, L. Dei, P.L. Nostro, A. Casnati, R. Gilles and P. Baglioni, Colloids Surf. A, 2000, 167 (1), 105.

137. P.Baglioni, E.R. Minten, and L. Kevan, J. Phys. Chem. 1988, 92, 4726.

138. H.J.D. McManus, Y.S. Kang, and L. Kevan, J. Phys. Chem. 1993, 97, 255.

139. H.A. Carime, J. Chem. Soc., Faraday Trans., 1998, 94, 2407.

140. P. Baglioni and L. Kevan, Prog. Colloid Polym. Sci., 1988, 76, 183.

141. M. Ravikant, D. Reddy and T.K. Chandrashekar, J. Chem. Soc. Dalton Trans., 1991, 2103.

142. P. Baglioni, A. Bencini, L. Dei, C.M.C. Gambi, P. Lo-Nostro, S.H. Chen, Y.C. Liu, J. Teixeira and L. Kevan, Colloids Surf. A, 1994, 88 (1), 59.

143. M. Ginley, U. Henriksson and P. Li, J. Phy. Chem., 1990, 94 (11), 4644.

144. T. Myassoedova, D. Grand and S. Hautecloque, Photochem. Photobiol. Sci., A, Chem., 1992, 64 (2), 159.

145. R. Gould, P.L. Kuo, and N.J. Turro, J. Phys. Chem. 1985, 89, 3030.

146. J. Nicholas and P.L. Kuo, J. Phys. Chem. 1986, 90, 837.

147. V.C. Reinsborough and V.C. Stephenson, Can. J. Chem. 2004, 82, 45.

148. J.C. Schulz and G.G. Warr, Ind. Eng. Chem. Res. 1998, 37, 2807.

149. E.M.E. Nemma and S.R. Salman, J. Incl. Phenom. Macro. Chem., 2004, 49, 267.

150. D.F. Evans, J.B. Evans, R. Sen, and G.G. Warr, J. Phys. Chem. 1988, 92, 784.

151. P.A. Quintela, R.C.S. Reno, and A.E. Kaifer, J. Phys. Chem. 1987, 91, 3582.

152. B. Sesta, A. Daprano, G. Maddalena and N. Proietti, Langmuir, 1995, 11 (8), 2860.

153. B. Sesta, A. Daprano, A. Princi, C. Filippi and M. Iammarino, J. Phys. Chem., 1992, 96 (23), 9545.

154. S. Ozeki, A. Kojima, S. Harada, S. Inokuma, H. Takahashi, T. Kuwamura, H. Uchiyama, M. Abe and K. Ogino, J. Phys. Chem., 1990, 94 (21), 8207.

155. S. Ozeki, A. Kojima, S. Inokuma, H. Takahashi, T. Kuwamura, H. Uchiyama, M. Abe and K. Ogino, J. Phys. Chem. (1990), 94 (21), 8213.

156. D.D. Miller, D.F. Evans, G.G. Warr, J.R. Bellare and B.W. Ninham, J. Colloid Interface Sci., 1987, 116 (2), 598.

157. J.M. Lehn and J.P. Sauvage, J. Am. Chem. Soc., 1975, 97 (23), 6700.

158. M.S. Bakshi, R. Crisantino, R. De-Lisi, and S. Milioto, Langmuir 1994, 10, 423.

159. Y. Moroi, E. Pramauro, M. Graetzel, E. Pelizzetti, P. Tundo, J. Colloid Interface Sci., 1979, 69 (2), 341.

160. S.D. Sandanayaka, Y. Araki, O. Ito, R. Chitta, S. Gadde and F. D'Souza, Chem. Commun., 2006, 4327.

161. A.D. Pidwell, S.R. Collinson, D.W. Bruce, S.J. Coles, M.B. Hursthouse and M. Schroder, Chem. Commun., 2000, 955.

162. O. Sutherland, J. Chem. Soc., Faraday Trans., 1986, 82 (1), 1145.

163. Y. Miyake, M. Yamada, T. Kikuchi and M. Teramoto, J. Chem. Soc., Faraday Trans., 1992, 88, 1285.

164. M.J. Wilson, R.A. Pethrick, D. Pugh and M.S. Islam, Chem. Soc., Faraday Trans., 1997, 93, 2097.

165. J. Nishimoto, E. Iwamoto, T. Fujiwara and T. Kumamaru, J. Chem. Soc., Faraday Trans., 1993, 89, 535.

166. F.P. Cavasino, C. Sbriziolo, M. Cusumano and A. Giannetto, J. Chem. Soc., Faraday Trans., 1989, 85 (1), 4237.

167. Y.C. Liu, P. Baglioni, J. Teixeira and S.H. Chen, J. Phys. Chem., 1994, 98 (40), 10208.

168. R.M. Izatt, K. Pawlak, J.S. Bradshaw and R.L. Bruening, Chem. Rev., 1991, 91 (8), 1721.

169. D.F. Evans, R. Sen and G.G. Warr, J. Phys. Chem., 1986, 90 (22), 5500.

170. S.A. Gerhardt, J.W. Lewis, J.Z. Zhang, R. Bonnett and K.A. McManus, Photochem. Photobiol. Sci., 2003, 2, 934.

171. G. Ilgenfritz, R. Schneider, E. Grell, E. Lewitzki, and H. Ruf Langmuir, 2004, 20, 1620.

172. M. Shamsipur and G. Khayatian, J. Incl. Phenom. Macro. Chem., 2001, 39, 109.

173. M.S. Bakshi, P. Kohli and G. Kaur, Bull. Chem. Soc. Jpn., 1998, 71, 1539.

174. M. Hasani and M. Shamsipur, J. Incl. Phenom. Macro. Chem., 1993, 16, 123.

175. R. Palepu and J.E. Richardson, Langmuir 1989, 5, 218.

176. M.S. Bakshi, J. Incl. Phenom. Macro. Chem., 1999, 33, 263.

177. M.S. Bakshi, J. Incl. Phenom. Macro. Chem., 2000, 36. 39.

178. D. Marji, K. Abbas, R. Saymeh and Z. Taha, J. Incl. Phenom. Macro. Chem., 1999, 34, 49.

179. T. Okubo, H. Kitano, and N. lse, J. Phys. Chem., 1976, 80 (24), 2661.

180. T. Okubo, Y. Maeda and H. Kitano, J. Phys. Chem. 1989, 93, 3721.

181. D.J. Jobe, R.E. Verrall, E. Junquera and E. Aicart, J. Phys. Chem. 1993, 97, 1243.

182. S. Milioto, M.S. Bakshi, R. Crisantino and R. De-Lisi, J. Solution Chem., 1995, 24 (2), 103.

183. E. Junquera, L. Pena, and E. Aicart, Langmuir, 1995, 11, 4685.

184. A.A. Rafatia, A. Bagheria, H. Iloukhania and M. Zarinehzad, J. Mol. Liq., 2005, 116, 37.

185. R. Palepu and V.C. Reinsborough, Can. J. Chem., 1988, 66, 325.

186. T. Okano, T. Tamura, T. Nakano, S. Ueda, S. Lee and G. Sugihara, Langmuir, 2000, 16, 3777.

187. S. Causi, R.D. Lisi, S. Milioto and N. Tirone, J. Phys. Chem. 1991, 95, 5664.

188. P.C. Shanks and E.L. Franses, J. Phys. Chem. 1992, 96, 1794.

189. J.J. Galan, A. Gonzalez-Perez, J.L. Del-Castillo and J.R. Rodriguez, J. Therm. Anal. Calorim., 2002, 70, 229.

190. J.J. Galan, A. Gonzalez-Perez, and J.R. Rodriguez, J. Therm. Anal. Calorim., 2003, 72, 465.

191. M.S. Bakshi, J. Colloid Interface Sci., 2000, 227, 78.

192. P. Baglioni, E.R. Minten, L. Kevan, J. Phys. Chem., 1988, 92, 4726.

193. M. Ravikant, D. Reddy and T.K. Chandrashekar, J. Chem. Soc., Dalton Trans., 1991, 2103.

194. V.C. Reinsborough and V.C. Stephenson, Can. J. Chem., 2004, 82, 45.

195. N.J. Turro and P.L. Kuo, J. Phys. Chem., 1987, 91, 3321.

196. P.P. Mishra, R. Adhikary, P. Lahiri and A. Datta, Photochem. Photobiol. Sci., 2006, 5, 741.

197. J. Gottstein and H. Scheer, Proc. Nati. Acad. Sci., 1983, 80, 2231.

198. A. Sokolowski, K.A. Wilk, U. Komorek, B. Rutkowski, L. Syper, Physicochem. Probl. Miner. Process., 2002, 36, 51.

199. Y. Zhang and Y.M. Lam, J. Nanosci. Nanotechnol., 2006, 6, 1.

200. Y.A. Gao, Z.N.Wang, J. Zhang, W.G. Hou, G.Z. Li, B. X. Han, F.F. Lui and G.Y. Zhang, Chin. Chem. Lett., 2005, 16 (7), 963.

201. B.H. Cipriano, S.R. Raghavan and P.M. McGuiggan, Colloid Surface Physicochem. Eng. Aspect, 2005, 262, 8.

202. K.D. Danov, S.D. Kralchevska, P.A. Kralchevsky, K.P. Ananthapadmanabhan, and A. Lips, *Langmuir,* 2004, **20,** 5445.

203. B.M. Razavizadeh, M.M. Khoshdel, H. Gharibi, R. B. Ardakani, S. Javadian, and B. Sohrabi , J. Colloid Interface Sci., 2004, 276, 197.

204. J. L. Qiao, W.J. Jin, C. Dong, and C. S. Liu, Chin. Chem. Lett., 1999, 10 (2), 161.

205. M.A. Muherei and R. Junin, J. Appl. Sci. Res., 2009, 5(2), 181.

206. M. J. Rosen, A. W. Cohen, M. Dahanayaki and X. Hua, J. Phys. Chem., 1982, 86, 541.

207. M. J. Rosen and S. Aronson, Coll. Surf., 1981, 3, 201.

208. S. Kaneshina, M. Manabe, G. Sugihara and M. Tanaka, Bull. Chem. Soc. Jpn., 1976, 49 (4), 876.

209. M. Manabe, K. Shirahama and M. Koda, Bull. Chem. Soc. Jpn.,1976, 49 (11), 2904

210. K. Fukada, J. Li, M. Fujii, T. Kato and T. Seimiya, J. Oleo. Sci., 2002, 51 (2), 103

211. J.G.A. Ramireza, V.V.A. Fernandeza, E.R. Maciasa, Y. Rharbib, P. Taboadac, R. Gamez-Corralesd, J.E. Puiga and J.F.A. Soltero, J. Colloid Interface Sci., 2009, 333 (2) 655

212. P. Morgado, R.Tomas, H. Zhao, M.C.D. Ramos, F. J. Blas, C. McCabe, and E.J.M. Filipe, J. Phys. Chem. C, 2007, 111, 15962

213. P.H. Stothart, Biochem. J., 1984, 219, 1049

214. C. Merino, E. Junquera, J. Jimenez-Barbero, and E. Aicart, Langmuir, 2000, 16, 1557

215. T.M. Letcher, J.D. M. Chalmers, and R.L. Kay, Pure & Appl. Chem., 1994, 66 (3), 419427

216. A. Gonzlez-Perez, J. M. Ruso, J. Nimo, and J. R. Rodriguez, J. Chem. Therm., 2003, 35 (12), 1983

217. S.B. Velasco, M. Turmine, D.D. Caprio, and P. Letellier, Colloids Surfaces A: Physicochem. Eng. Aspects, 2006, 275 (1-3), 50

218. P. Brocos, N. Daz-Vergara, X. Banquy, S. Prez-Casas, M. Costas, and N. Pieiro, J. Phys. Chem. B, 2010, 114 (39), 12455

219. C. Cabaleiro-Lago, L. García-Ro, P. Hervs and J. Prez-Juste , J. Phys. Chem. B, 2009, 113 (19), 6749

220. L. Li, X. Guo, L. Fu, R.K. Prud'homme and S.F. Lincoln, Langmuir, 2008, 24 (15), 8290

221. S. Talwar, J. Harding, K.R. Oleson and S.A. Khan, Langmuir, 2009, 25 (2), 794

222. D. Taura, A. Hashidzume, Y. Okumura and A. Harada, Macromolecules, 2008, 41 (10), 3640

223. M. Miyake, K. Yamada and N. Oyama, Langmuir, 2008, 24 (16), 8527

224. K. Szymczyk and B. Jaczuk, Langmuir, 2009, 25 (8), 4377

225. M.P. Cashion, X. Li, Y. Geng, M.T. Hunley and T.E. Long ,Langmuir, 2010, 26 (2), 678

226. C. Das and B. Das, J. Chem. Eng. Data, 2009, 54 (2), 559

227. K. Din, M.S. Sheikh, and A.A. Dar, J. Phys. Chem. B, 2010, 114 (18), 6023.